TM

ROBOTICS
CAREERS
Preparing for the Future

SIMONE PAYMENT

rosen publishing's
rosen central

NEW YORK

Published in 2011 by The Rosen Publishing Group, Inc.
29 East 21st Street, New York, NY 10010

Library of Congress Cataloging-in-Publication Data

Payment, Simone.
Robotics careers: preparing for the future / Simone Payment.
   p. cm. — (Robotics)
Includes bibliographical references and index.
ISBN 978-1-4488-1239-4 (library binding)
ISBN 978-1-4488-2253-9 (pbk.)
ISBN 978-1-4488-2254-6 (6-pack)
1. Robotics–Vocational guidance–Juvenile literature. I. Title.
TJ211.25.P39 2011
629.8'92023–dc22

                                             2010024134

Manufactured in the United States of America

CPSIA Compliance Information: Batch #W11YA: For further information, contact Rosen Publishing, New York, New York,
at 1-800-237-9932.

On the cover: DynamicBrain (DB), a human-shaped robot, displays its
ability to replicate human behavior.

CONTENTS

INTRODUCTION

Robots are more popular than ever before. All around the world, teams of engineers, computer programmers, technicians, and researchers are working together to create new robots, improve existing robots, and discover new applications for robots. In hospitals all over the country, robots help surgeons perform advanced surgery. Robots are used to explore places that are hostile to human beings, such as deep in the oceans or on the surface of other planets. In the military, robots are used to scout an area to make sure it is safe before troops enter. Thousands of factories use robots to make cars, pack boxes, and handle chemicals.

In schools all over the United States, students are joining robotics teams and entering robotics competitions. It's common for students gearing up for a competition to spend long hours working on robots and troubleshooting problems that arise. Robotics competitions, such as those held by FIRST (For Inspiration in Science and Technology) and BEST (Boosting Engineering Science and Technology) are massive, thrilling events where hundreds of robotics teams comprised of thousands of students compete against one another.

The field of robotics thrives on young people's innovation and creativity. Students who are members of robotics teams today may be the leaders of the robotics industry tomorrow. Being on a robotics team is fun, but it's also hard work. In fact, robotics teams help prepare students for the kind of intense working environment they would find as professional roboticists. Much like professional roboticists, members of robotics teams must learn how to tackle technical problems, communicate with people who have different interests or specialties in order to complete a project, and come up with solutions to problems under a deadline.

Every year, new applications for robots are being developed, new jobs associated with robotics are being created, and great advances in robots are being made. We live in a time when high school students have access to tools that will allow them to build sophisticated robots, college students work alongside professional roboticists on high-profile robotics projects, and simple robots are commercially available to the general public. Despite the advances that have been made in robotics, it is only the tip of the iceberg. For robotics researchers, computer scientists, robotics engineers, and robotics technicians, the future is bright—and there is much work to be done.

CHAPTER 1
Robotics Researche

Robotics researchers are the people driving amaz
ing new developments in the field of robotic
They work on ways to make robots functio
better and find new practical applications for robot
Robotics researchers need to be flexible and creative
since they often set out to solve problems that no on
has ever tackled before.

Some researchers work in the general field of robo
ics, and others specialize in a particular area. This mean
that a robotics researcher might specialize in applyin
his or her knowledge of a specific science, such as bio
ogy, to robotics. A researcher with a specialization i
biology might work on creating lifelike prosthetic limb
or finding ways to create robots modeled on biologic
designs. For instance, robotics researcher Sangbae Kin
who is a robot designer and an assistant professor c
mechanical engineering at the Massachusetts Institut
of Technology (MIT), studied geckos, which are able t
climb walls. Geckos have tiny hairs on their feet that ac
almost like suction cups. Kim used this idea to desig
an extremely effective robot that is able to climb wall
even those with perfectly smooth surfaces.

Researchers at universities in Switzerland and Franc
studied how salamanders move, especially how salaman
ders' four feet move independently and how they switc

to a different kind of movement when leaving land and entering water. This helped them figure out ways to make robots with four feet move more easily over rough surfaces and led to the development of a robot known as Salamandra Robotica. Built by the Biologically Inspired Robots Group (BIRG), Salamandra Robotica has four feet and a body divided into nine segments. Each segment is powered by a separate microcontroller.

Some researchers radically rethink the way that robots could possibly function. For instance, robotics researcher Rodney Brooks has worked on creating robots that can figure out how to interact with their surroundings. Brooks, along with many students he has taught, is responsible for numerous advances in the field of robotics. He has built many robots, including buglike robots and humanoid robots, and is one of the leading researchers in the field of artificial intelligence (AI). He built small, simple, inexpensive robots based on insects that could navigate their environment. Brooks also worked on developing a robot called Cog, which could interact with human beings, and oversaw the development of Kismet, a robot that has the ability not only to interact with people but also read and display facial expressions.

The work of robotics researchers is often very theoretical but generally has a practical application. The possibilities inherent in the world of robotics are limitless, and robotics researchers are the people who come up with the cutting-edge robots of the future.

RODNEY BROOKS: ROBOTICS RESEARCHER

RODNEY BROOKS IS ONE OF THE WORLD'S BEST-KNOWN ROBOTICS RESEARCHERS. HE HAS TAUGHT AND DONE RESEARCH AT CARNEGIE MELLON UNIVERSITY, STANFORD UNIVERSITY, AND MIT. HE HAS ALSO FOUNDED TWO ROBOTICS COMPANIES: IROBOT IN 1990 AND HEARTLAND ROBOTICS IN 2008. PERHAPS MOST FAMOUSLY, IROBOT MAKES THE ROOMBA, A ROBOTIC VACUUM CLEANER, AND IT MAKES A NUMBER OF OTHER ROBOTS FOR HOME USE. IT ALSO MAKES MILITARY ROBOTS SUCH AS THE PACKBOT, A MOBILE ROBOT THAT CAN SCOUT LOCATIONS, AND EDUCATIONAL ROBOTS FOR STUDENTS AND TEACHERS. HEARTLAND ROBOTICS IS DEDICATED TO INTEGRATING ROBOTS INTO THE AMERICAN WORKFORCE.

At a conference in India, Rodney Brooks introduced a student competition to design a human settlement on the moon.

BROOKS IS INTERESTED IN FINDING NEW WAYS OF THINKING ABOUT ROBOTS AND MAKING ROBOTS THAT ORDINARY PEOPLE CAN INTERACT WITH. OBSESSED WITH ROBOTS FROM THE TIME HE WAS A YOUNG BOY, BROOKS WAS ESPECIALLY INTERESTED IN CREATING "INTELLIGENT" ROBOTS. HE CREATED A SIX-LEGGED, INSECTLIKE ROBOT NAMED GENGHIS IN THE MID-1980S. PROPOSED FOR USE IN SPACE MISSIONS, ROBOTS LIKE GENGHIS HAD THE ADVANTAGE OF BEING SMALL, INEXPENSIVE, AND ABLE TO OPERATE AUTONOMOUSLY. A LARGE NUMBER OF GENGHIS-LIKE ROBOTS COULD EXPLORE A HOSTILE ENVIRONMENT LIKE THE SURFACE OF ANOTHER PLANET, AND THE MISSION WOULD NOT BE JEOPARDIZED IF A FEW OF THEM WERE DESTROYED. BROOKS WAS FEATURED IN THE 1997 DOCUMENTARY *FAST, CHEAP, AND OUT OF CONTROL*. DIRECTED BY ERROL MORRIS, THE FILM FEATURED FOUR PEOPLE WITH INTERESTING CAREERS.

WORKING AS A RESEARCHER

Robotics researchers are often professors at colleges and universities. An academic environment provides them with the logistical and financial support they need to conduct their research and allows them to pass on their specialized knowledge to their students in the classroom. As professors, they might lecture to large groups of students or teach smaller classes. They will also spend time working individually with upper-level students who are working on their graduate degrees.

Robotics researchers sometimes work alone, but more often they form part of a larger team. They may work with robotics engineers, computer programmers, robotics technicians, or other scientists. They generally work in laboratories or at testing sites. Often able to set their own schedules, they can work nights or weekends if they choose. Robotics researchers who work at colleges or universities sometimes have summers off, at least from their teaching duties. Some choose to use this free time as a vacation, while others spend it doing further research, often with the assistance of university students. Researchers working in academia often publish their findings in academic journals.

PRIVATE ROBOTICS COMPANIES

Some robotics researchers work for private robotics companies. At a robotics company, researchers might work on a specific project, such as finding a way to create a more efficient robotic crop harvester. Or they might work on a more general project that could have many applications, such as devising ways to improve facial recognition software to allow robots to tell the difference between individual human faces. Other robotics researchers work for the military or defense contractors to design unmanned vehicles, or for government agencies, such as NASA, designing robots that work on the

International Space Station. The salaries that robotics researchers earn can vary greatly depending on their experience and whether they are employed by a university or a private company.

REQUIREMENTS

Like anyone working in the field of robotics, robotics researchers will use lots of science and math in their work, so strong skills in those areas are necessary. Applying oneself to mastering those subjects in high school and college is key. It's also a good idea for students to take drafting or electronics classes, and to study engineering to learn about the practical concerns that go into building a robot.

Some robotics researchers major in computer science, electrical or mechanical engineering, or even disciplines such as aeronautics in college before eventually turning their attention to robotics research. Successful robotics projects often draw on the diverse talents of a group of people who are experts in different disciplines, so having a wide base of knowledge is key.

EDUCATION

Robotics researchers require many years of education. They generally begin their career by getting a four-year bachelor's degree, usually in a branch of science, engineering, or computing. Then they go on to earn a master's degree or a doctorate. Most robotics researchers who work as professors have at least a master's degree as well as additional postgraduate degrees.

It takes six years or more to earn a doctorate degree. While completing master's or doctorate programs, students work with professors and other students on research projects. Students can usually choose their own area of study and are supervised by a faculty member with similar interests. To complete the degree, students must write a dissertation that describes their research or give an oral presentation. With a bachelor's or master's degree, it's possible to work for robotics companies or for the government. To get a job as a robotics researcher at a university, it is usually necessary to have a doctorate.

COMMUNICATION SKILLS

Being a skilled communicator is especially useful for robotics researchers. Besides being useful when teaching students, good communication skills can help researchers express their ideas to the people they are collaborating with. Robotics researchers are often called upon to solve unique problems, so being able to think critically and come up with creative and practical solutions is an important skill. Robotics researchers must come up with research ideas on their own, so they must be independent and able to motivate themselves.

An interest in learning is also useful for robotics researchers because most continue their education throughout their lifetimes. Beyond the basic educational requirements that are necessary to become a professional roboticist, robotics researchers should actively develop interests in other areas. Subjects such as history and biology might not seem to relate to robotics, but they can help students learn to think outside the box.

BUILDING ROBOTICS SKILLS NOW

It's never too early to start preparing for a career in robotics. Many young people build their own robots using a kit, such as the LEGO Mindstorms kit. Students should join their school's robotics team. If their school doesn't have a robotics

This robotics team prepares for a FIRST competition in Washington, D.C. Coaches from the Office of Naval Research sponsor this team, sharing their robotics knowledge and skills.

team, they should consider talking to school administrators and teachers about starting one. Some colleges and robotics companies sponsor high school robotics teams.

There are also many activities outside of school that can be useful for preparing for a robotics career. Reading up on robots in the school library or online is a good place to start. There are many robotics summer camps in the United States that students can attend and learn valuable skills while having a great time.

CHAPTER 2
Computer Scientist

Robots perform the tasks that they are programmed to do. For instance, factory robots are programmed to repeat certain actions when manufacturing a product. As computer technology and programming have grown more sophisticated, so have robots. Today's robots can do more than simply perform a single task over and over again—instead, they can interact with their environment, respond to human beings, and move independently.

Computer scientists design and write the software that allows robots to perform functions. They might work on devising new programs and programming languages, or they might work on advanced research projects. Some computer scientists program individual robots or design a system to run a particular kind of robot. These computer systems might be built into the robot itself, or they might be part of a controller that operates the robot.

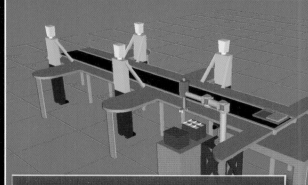

RoboWorks software helps computer programmers create 3-D models of manufacturing and other industrial robots. The models help programmers test robots.

AARON EDSINGER: ROBOTICIST

AARON EDSINGER IS A POSTDOCTORAL ASSOCIATE AT MIT'S COMPUTER SCIENCE AND ARTIFICIAL INTELLIGENCE LABORATORY. ALTHOUGH HE STARTED OUT AS AN ARTIST, EDSINGER GOT INTERESTED IN BUILDING ROBOTS AND THE SCIENCE OF ROBOTICS. HE ATTENDED STANFORD UNIVERSITY IN CALIFORNIA, WHERE HE EARNED A BACHELOR OF SCIENCE DEGREE IN COMPUTER SYSTEMS AND ENGINEERING. EDSINGER GOT A MASTER'S DEGREE—AND THEN A PH.D.—IN COMPUTER SCIENCE FROM MIT. EDSINGER'S ADVISER AT MIT WAS ROBOTICS RESEARCHER RODNEY BROOKS. WHILE IN COLLEGE, EDSINGER WORKED AS BOTH A RESEARCH ASSISTANT AND A TEACHING ASSISTANT.

FASCINATED BY THE IDEA OF CREATING ROBOTS THAT COULD HELP PEOPLE, EDSINGER AND HIS COLLABORATOR, JEFF WEBER, BUILT DOMO—A HUMANOID ROBOT THAT CAN SAFELY INTERACT WITH PEOPLE. DOMO IS EQUIPPED WITH A NUMBER OF SENSORS THAT ALLOW IT TO SENSE ITS ENVIRONMENT AND CAMERAS THAT ALLOW IT TO HAVE VISUAL PERCEPTION. IF HANDED AN OBJECT, DOMO CAN HOLD IT UP AND FIGURE OUT ITS DIMENSIONS. BY INTERACTING WITH PEOPLE AND THE WORLD AROUND IT, DOMO CAN ACTUALLY "LEARN" AND ADAPT TO ITS SURROUNDINGS. ROBOTS LIKE DOMO MIGHT SOMEDAY BE ABLE TO ASSIST PEOPLE WITH LIMITED OR IMPAIRED MOBILITY, SUCH AS THE ELDERLY.

EDSINGER HAS COFOUNDED TWO COMPANIES: HEEHEEHEE LABS AND MEKA ROBOTICS. HE BELIEVES THAT PATIENCE IS KEY TO BEING A GOOD ROBOTICIST, AS DEVELOPING A ROBOT CAN TAKE YEARS. ROBOTICISTS OFTEN HAVE TO START OVER FROM SCRATCH, AND BECAUSE OF THIS, EDSINGER BELIEVES IT'S ALSO IMPORTANT FOR ASPIRING ROBOTICISTS TO LOVE WHAT THEY DO.

PROGRAMMING ROBOTS

When computer scientists are designing a program for a robot, they usually develop a computer model first. This computer model simulates both the robot and the program they are writing to control it. They continue to change and improve the program until the on-screen robot works correctly. This can take some time, and it involves a lot of trial and error. Computer programs rarely work correctly right from the start.

Aaron Edsinger puts objects in a box held out to him by robot Domo, in his lab at MIT.

FIRST ROBOTICS COMPETITIONS

EACH YEAR, FIRST HOLDS NATIONAL ROBOTICS COMPE-TITIONS. ROBOTICS TEAMS RECEIVE KITS CONTAINING PARTS AND INSTRUCTIONS FROM FIRST. THEY THEN WORK TOGETHER, UNDER A DEADLINE, TO BUILD ROBOTS AND COMPETE IN LOCAL COMPETITIONS. WINNERS FROM LOCAL EVENTS ADVANCE TO A NATIONAL TOURNAMENT. THERE THEY COMPETE AGAINST STUDENTS FROM ALL OVER THE UNITED STATES AND TWELVE OTHER COUNTRIES. ALTHOUGH THE MAIN COMPETITION IS FOR STUDENTS IN GRADES 9–12, THERE ARE SEPARATE COMPETITIONS FOR STUDENTS IN GRADES 4–8. STUDENTS AS YOUNG AS SIX CAN COMPETE.

It's common for computer programmers to write and rewrite computer code dozens—or even hundreds—of times. Once the computer code is working properly, it is loaded into the actual robot.

Computer scientists need to be able to think logically, have a good grasp of cause and effect, be able to break complex tasks down into smaller steps, and be able to handle multiple projects simultaneously. Computer scientists usually work in laboratories or offices, although some are able to work from home. They often work on a team with robotics engineers and technicians. They must be able to work well independently and work on a team with others. To find out the salaries that computer scientists can expect to earn—and salaries for the rest of the careers in this book—please consult the U.S. Bureau of Labor Statistics' *Occupational Outlook Handbook* (http://www.bls.gov/oco).

PREPARING FOR A CAREER AS A COMPUTER SCIENTIST

Students interested in becoming computer scientists need to be good at math and science. They should pursue their passion for computers by taking computer classes in high school and working on computer projects outside of school. This can involve joining a computer club, building a Web page, or even learning how to make an iPhone app. Obviously, students will also want to be familiar with robotics and should join a robotics team and take robotics classes in college.

To become a computer scientist, it is necessary to have at least an associate's degree in computer science or computer engineering. Most computer scientists and programmers have a bachelor's degree, and many go on to get a master's degree or a doctorate. While working toward a graduate degree, students do research projects in their area of interest. Many degree programs require that students write research papers or give presentations about their findings.

Universities with robotics programs offer various areas in which students can specialize. For example, at Stanford University's computer science graduate program, students can choose from a variety of computer-related specialties, including artificial intelligence. In that program, students can study how computer science is used in robotics, how machines "learn," and how machines process language.

Some computer science programs offer internships during the semester or help students find internships during the summer. Internships can give students a better idea of what it is like to work as a computer programmer, and working in a professional environment allows students to hone their skills and gain practical experience. Internships can also provide students with a better idea of what kind of tasks they like and where their skills lie. Many colleges help students find jobs at local companies.

CHAPTER 3
Robotics Engineer and Robotics Technician

Robotics engineers design, build, program, and test robots and robotic devices. They solve technical problems such as how to get a robot to perform a particular task and analyze the best way to find creative solutions to robotics problems.

Robots are extremely complex machines, so it's common for engineers who specialize in different fields of engineering to work together when creating them.

Robotics engineers generally focus on electrical engineering or mechanical engineering. Electrical engineers specialize in designing and developing electrical systems. Electrical engineers who specialize in robotics might design the electronics that control the robot, the robot's electronic sensors, and other electrical components.

Mechanical engineers specialize in the design and development of mechanical systems. They might design the components in a robot's "hand" that allow it to pick up an object. Or, they might help develop the components that allow a robot to walk. Industrial and mechanical engineers might also work in robotics. These types of engineers focus on what's involved in the manufacturing process.

ROBIN MURPHY: ROBOTICS ENGINEER

ROBIN MURPHY IS A ROBOTICS EXPERT WHO HAS DESIGNED ROBOTIC DEVICES THAT SEARCH FOR VICTIMS IN DISASTER AREAS OR IN COLLAPSED BUILDINGS. THE ROBOTS ARE SMALL ENOUGH TO ENTER SPACES WHERE HUMANS WOULDN'T FIT AND STURDY ENOUGH TO WITHSTAND DANGEROUS CONDITIONS. THEY ARE ALSO SMART ENOUGH TO OPERATE SEMI-INDEPENDENTLY OF THEIR HUMAN CONTROLLERS AND MAKE SOME OF THEIR OWN DECISIONS. A MECHANICAL ENGINEER WITH A PH.D. IN COMPUTER SCIENCE, MURPHY HAD PLANNED TO SPECIALIZE IN SPACE EXPLORATION UNTIL SHE REALIZED THAT SHE COULD MAKE A BIGGER DIFFERENCE IN PEOPLE'S LIVES BY DESIGNING RESCUE ROBOTS.

MURPHY AND HER ROBOTS WORKED AT GROUND ZERO FOLLOWING THE ATTACKS ON THE WORLD TRADE CENTER IN NEW YORK CITY ON SEPTEMBER 11, 2001; THEY WORKED IN NEW ORLEANS AFTER HURRICANE KATRINA; AND THEY CONTINUE TO HELP OUT AT MANY OTHER DISASTER SITES. MURPHY IS A PROFESSOR OF COMPUTER SCIENCE AND ENGINEERING AT TEXAS A&M UNIVERSITY.

WORKING AS A ROBOTICS ENGINEER

Robotics engineers may focus on a particular industry, such as manufacturing. They might study the automotive industry extensively and then design robots to efficiently perform manufacturing tasks. This engineer might spend months or years designing a robotic system and then updating and revising it when it is put into use. Others may work for a company that uses robots. For example, a robotics engineer might work in an automotive plant designing the robots that work on the assembly line.

Robotics engineers are frequently called upon to work on-site. They might go to a manufacturing plant to upgrade robots or a hospital to observe how surgeons are using robotic devices they have designed. Some robotics engineers work for a company that specializes in designing and building robots. The amount of money that robotics engineers can expect to make largely depends on what kind of company they are employed by.

PREPARING FOR A CAREER AS A ROBOTICS ENGINEER

Robotics engineers need to be good at math and science, and skilled at working with their hands. They also need to be creative thinkers who enjoy coming up with innovative solutions to problems and puzzles. Robotics engineers must be disciplined and capable of motivating themselves to do a good job in a timely manner. As part of their job, they are constantly learning new things and keeping up to date with the latest developments in the field.

Students interested in becoming robotics engineers should take as many algebra, geometry, and calculus classes as possible. They should take classes in physics and chemistry, and consider taking classes in automotive mechanics, drafting, computer-aided design (CAD), and electronics. Aspiring robotics engineers that work hard can go from building robots from kits to designing robots as part of a design team.

In college, it's a good idea to explore many different kinds of engineering. Classes in computer programming are essential, as robotics engineers often

This electrical engineering major is soldering a circuit board. Along with three classmates, she built a robot, assembling and installing its electrical components.

need to do some programming or at least be able to work closely with computer programmers when designing or building robots. The field of robotics changes quickly, so engineers must always be sure to stay on top of the newest technology.

Robotics engineers need to earn a bachelor of science degree. Some colleges offer a specific degree in robotics engineering. However, most robotics engineers get a degree in another type of engineering, such as mechanical engineering or electrical engineering. While working toward that degree, they take courses in robotics. Most robotics engineers continue on to get a master's degree, and others continue on to get a doctorate. These degrees can help a robotics engineer advance his or her career. Some universities offer advanced degrees in robotics. For example, the University of Texas at Austin's Robotics Research Group offers master's and doctorate degrees in mechanical engineering with a focus on robotics.

ROBOTICS TECHNICIANS

Robotics engineers often work alongside robotics technicians. Robotics technicians perform practical, hands-on robotics work. They build the robots and robot systems that engineers design, and repair them when they break down. They also test and maintain robots and robotic parts so that they will continue to work smoothly.

Robotics technicians usually work on-site. Sometimes the work site can be dangerous, so technicians have to

Researchers at Carnegie Mellon University's Robotics Institute work on Qwerkbot, a three-wheeled robot.

be careful to observe the proper safety precautions. They may work closely with the robotics engineers who design the robots and the computer scientists who program them. They install robots in factories and set up robotic systems in laboratories. If the robots used in a manufacturing plant need to perform new tasks, robotics technicians are usually the people who make the necessary adjustments that allow this to happen. Robotics technicians may also train people who use robots. When a new robot or robotic system is installed in a factory, technicians would teach workers how to use and safely interact with the robot.

PREPARING FOR A CAREER AS A ROBOTICS TECHNICIAN

Robotics technicians must be mechanically minded, and they should excel at using tools. Students who would like to be robotics technicians should take classes in math, automotive mechanics, CAD, drafting, electronics, and other technical classes. Good math and problem-solving skills are essential for robotics technicians.

Most robotics technicians get an associate's degree. Others choose to get a bachelor's degree in robotics or a related field, such as computer programming or engineering. It is always a good idea to continue learning with on-the-job training or seminars.

Robotics technicians can sometimes become specialists in more than one field. For example, a robotics technician that works with automated welding systems might get training in welding as well as in robotics. In order to help program and maintain the automatic welders, the technician must know the basics of how to weld and how non-robotic welding systems work. This additional training helps the technician in his or her current job. It also allows a technician

Any robotics career will require a good command of math, so it's a good idea to take as many math classes as you can fit into your school schedule.

to expand his or her knowledge base for future job opportunities.

Robotics technicians might also want to become certified as manufacturing technologists—people who analyze how robots and other manufacturing machines can do their jobs more efficiently within the workplace. The Society of Manufacturing Engineers offers a program that allows technicians to become certified manufacturing technologists. This involves passing a test after fulfilling school and work requirements.

CHAPTER 4
Robotics Applications

G rowth in the robotics industry is expected to continue at a fast pace in the years to come. This chapter will cover some of the many industries that use robots. People pursuing a career in robotics might find themselves working as part of a team engaged in creating or maintaining robots used in one of these industries.

HEALTH CARE AND MEDICAL

Surgeons use robots to assist them in performing many types of surgery. These surgical robots can perform precise, delicate operations and have better fine motor control than a human surgeon. Robotic surgery is also usually less invasive than traditional surgery, allowing patients to heal more quickly. One of the most popular surgical robots is the da Vinci Surgical System, which is built by the company Intuitive Surgical, Inc. When using the system, the surgeon does not operate directly on the patient. Instead, he or she controls the robot while sitting at a computer console and watching the surgery take place on a screen in real time.

Robots are also used in pharmacies to fill prescriptions. Robotic systems locate the requested drug, dispense the correct amount of medication, and then fill and label a container. Pharmacy robots are extremely accurate and cut down on medication errors.

Robotics researchers are making many advances in improving the connection between a person's brain and a robotic limb. They hope to create advanced artificial limbs and other body parts that can function nearly as well as human limbs. They are also working on robots than can help nurses perform their daily tasks, as well as robots that could aid people living in nursing homes or who need care in their own homes.

MILITARY

The military uses robots for a number of different purposes and is constantly looking for ways to apply robotics to military applications. For instance, the military sometimes uses vehicular robots to go into dangerous areas ahead of ground troops. Some of these advanced robots are

Bigdog image courtesy of Boston Dynamics © 2010

Big Dog, a robot designed for military applications, can easily navigate rough or rocky terrain—even an ice- and snow-covered hill.

vehicles, such as the tactical unmanned ground vehicle (TUGV). Humans operate the TUGV remotely. A TUGV can enter an area before troops to check out the situation on the ground. The military also uses robotic unmanned planes. Controlled from the ground, these planes fly high over an area to perform a survey or a search. They send high-quality images back to controllers on the ground.

One experimental military robot is called Big Dog. Big Dog is a robot that can assist human soldiers with a number of tasks. Walking on four legs, it can carry up to 400 pounds (181 kilograms) of equipment and cargo. It can travel up to 20 miles (32 kilometers) over rough ground and go places many vehicles cannot, such as tree-covered hillsides. Big Dog can even run and faithfully follow a human leader. Scientists at Boston Dynamics are working with the Defense Advanced Research Projects Agency (DARPA), a U.S. government agency that funds technological research on behalf of the U.S. Department of Defense. Together, Boston Dynamics and DARPA are working to make a version of the robot that can perform additional tasks, like recharging batteries. Other military robots are able to detect or blow up bombs. Some police and fire departments also use this type of robot.

AEROSPACE

Aeronautics and space programs around the world rely on countless robots. For instance, the International

Space Station (ISS) has a robotic arm that lifts and moves equipment and cargo. The ISS also has a robot named Dextre, which can work inside the station or attach to the end of the robotic arm outside the station. Dextre can fix or replace broken parts and move small loads around the station. One major advantage of using robots at the ISS or on other space missions is that they don't need special equipment—like astronauts do—to work in the oxygen-free atmosphere of space.

Another type of space robot is a rover. Mars Exploration Rovers (MERs) have been roaming the surface of Mars for several years. Robotics engineers on Earth control the MERs as they collect samples of the Martian soil and rocks. Onboard computers analyze the samples and send data back to Earth.

This launch vehicle, under construction at Kennedy Space Center in Orlando, Florida, may someday launch the ARES robotic plane.

THE ASTRONAUTS' SIDEKICK

ROBONAUT 2, OR R2, IS A ROBOT MADE BY NASA AND GENERAL MOTORS (GM). THE ORIGINAL VERSION OF ROBONAUT WAS DESIGNED TO GO ON SPACE MISSIONS WITH ASTRONAUTS. R2, A MORE RECENT VERSION, IS CONSIDERED TO BE ONE OF THE MOST ADVANCED HUMANOID ROBOTS EVER MADE.

ABOUT THE SIZE OF A HUMAN BEING, IT HAS HANDS THAT CAN GRIP AND HOLD OBJECTS. IT IS ABLE TO USE THE SAME TOOLS THAT PEOPLE DO. R2 ALSO HAS CUTTING-EDGE SENSORS THAT ALLOW IT TO TELL WHERE HUMANS AND OTHER OBJECTS ARE LOCATED.

NASA AND GM DESIGNED R2 WITH SAFETY IN MIND BECAUSE R2 MAY NEED TO WORK CLOSELY WITH HUMANS IN DANGEROUS SITUATIONS.

The National Aeronautics and Space Administration (NASA) is currently developing ARES, a robotic plane that may someday explore Mars. ARES is designed to parachute from a spacecraft and unfold as it falls. About 1 mile (1.6 km) from the surface, ARES will begin soaring over the Martian landscape. This will allow it to take photographs and collect other data.

LOOKING FORWARD: ROBOT ETHICS

FROM THE EARLY DAYS OF ROBOTICS, HUMANS HAVE CON-SIDERED THE ETHICS OF BUILDING AND USING ROBOTS. ARE THERE TYPES OF ROBOTS HUMANS SHOULD NOT DESIGN OR BUILD? SHOULD ROBOTS BE ALLOWED TO REPRODUCE THEM-SELVES? IF ARTIFICIAL INTELLIGENCE BECOMES ADVANCED ENOUGH TO PRODUCE VERY INTELLIGENT ROBOTS, WILL THEY HAVE RIGHTS? SHOULD ROBOTS BE ALLOWED TO MAKE DECI-SIONS, SUCH AS WHETHER OR NOT TO FIRE A WEAPON? WHO WOULD BE RESPONSIBLE IF A ROBOT HURTS SOMEONE? WHAT KIND OF ETHICS SHOULD GUIDE HUMAN BEINGS' TREATMENT OF ROBOTS? DECISIONS WILL HAVE TO BE MADE ABOUT MANY OF THESE QUESTIONS, AND MANY OTHER QUESTIONS WE HAVE NOT YET EVEN THOUGHT OF. IN THE FUTURE, THERE MAY BE PEOPLE WHO SPECIALIZE IN THE ETHICS OF ROBOTICS.

FACTORY AUTOMATION

More than 150,000 robots are used in factories world-wide. A robot named Unimate was the first industrial robot. Built in 1961, it had one arm that could lift as much as two tons at a time. Unimate's job was to stack metal parts.

Today, robots used in factories are much more complex and are capable of many tasks. They perform duties such as lifting, stacking, welding, and packing. Some robots are used in factories to perform delicate jobs, like packaging cookies or candy on an assembly line. Others lift and position heavy car parts in automobile factories. Factory robots can also attach metal parts together and weld them faster and more accurately than human welders. Robots can perform the same task over and over without making mistakes, getting tired, or getting bored. They are also cleaner than humans, which can be important for types of manufacturing that require a dust-free or sterile environment. Robots also don't need to be paid a salary, and they never need to stop to work or sleep. This can allow some factories to run for twenty-four hours a day.

They can do jobs that are too dangerous, tedious, or dirty for humans. Some robots, such as the Envirobot, manufactured by Chariot Robotics, can work with toxic chemicals and other substances. The Envirobot can quickly remove paint from steel surfaces, such as the hulls of ships, in an environmentally safe say.

ARTIFICIAL INTELLIGENCE

One of the most exciting areas for new developments in robotics is artificial intelligence (AI). Scientists involved in the field of AI research ways to create robots with the ability to learn, solve problems, and even reason. It's possible that someday robots with AI will learn from their experiences, just like human beings. Robots could also make their own decisions based on data they have collected.

An example of this is the Autonomous Loading System (ALS) developed at the National Robotics Engineering Center at Carnegie Mellon University's Robotics Institute. The ALS is a robotic excavator that can dig up material in mines or excavation sites. It then loads trucks with excavated soil or rocks. The ALS decides how and where to dig based on what type of material it will be digging up. It then figures out the fastest and easiest path to get the excavated material into a waiting truck. The ALS scans the area to find the truck and avoid any obstacles in its path.

Another example of a robot with AI is Adam, a "robot scientist." Adam was designed by a research group at Aberystwyth University in the United Kingdom, and it is able to conduct science experiments independently. Adam performs the first step of an experiment and then records the results. Using data from the first experiment, it is able to figure out what experiment to perform next. Data from the second experiment leads Adam to the next step, and so on. It

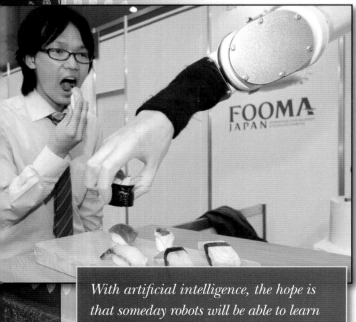

With artificial intelligence, the hope is that someday robots will be able to learn from their experiences.

has already made a scientific breakthrough—it found a solution to a genetics question scientists had been working to solve for years. After Adam's experiments answered this genetics question, human researchers confirmed Adam's findings.

Artificial intelligence is a rapidly developing field, and its impact on the future of robots will be substantial. In the years to come, advances in robotics are also sure to continue at a fast pace. The U.S. Department of Labor's Bureau of Labor Statistics, a government agency that tracks job trends, expects that careers associated with robotics will continue to grow in the coming years, making robotics an excellent career choice.

Adam, the robot scientist, carries out lab experiments but also can think and reason enough to plan its next experiment on its own.

INTERVIEW WITH
GAIL DRAKE

Gail Drake is a teacher at Battlefield High School in Haymarket, Virginia. After working as a college professor and IT professional, Drake pursued a teaching career and started a FIRST robotics team at her school. Here's what she had to say:

The school board had a request from a local corporation for the district to start a FIRST robotics team, and a school board member asked me to start the team. I talked with my colleagues, one of whom had been a member of a team in college. Four of us attended a meeting with Dave Laverty from NASA to learn about FIRST. In six weeks we started a team, and the game rules were announced on the first Saturday of the calendar year. The build season was six weeks. Those were the craziest six weeks of my life.

I do not know of any other team sport/activity that is growing at the same rate as robotics. Robotics teams have experienced a 7,500 percent growth rate since 2005, or an average growth rate of 1,500 percent per year.

HOW DOES BEING A MEMBER OF A FIRST ROBOTICS TEAM PREPARE ONE FOR A CAREER AS A PROFESSIONAL ROBOTICIST?

There is a direct parallel from FIRST robotics teams to corporate and military robotics. Both require

FAMILIARITY OF MECHANICAL AND ELECTRICAL ENGINEERING, PROGRAMMING, CAD, BUILDING SKILLS, AND PROBLEM SOLVING. THE STRENGTH OF YOUR FRC WILL DETERMINE THE JOB FUNCTIONS THAT ARE TAUGHT AND PRACTICED ON THE TEAM.

DO COLLEGE RECRUITERS—AND LATER, POTENTIAL EMPLOYERS—LOOK FOR PEOPLE WITH A WIDE BASE OF EXPERIENCE?

IT IS IMPORTANT FOR YOUNG PEOPLE TO TRY NEW THINGS AND DISCOVER WHERE THEIR INTERESTS LIE. ROBOTICS IS A WIDE-REACHING FIELD THAT ALREADY ENCOMPASSES MANY DIFFERENT SUBJECTS. IT INVOLVES MORE THAN SIMPLY THE ACT OF BUILDING A ROBOT, AND COLLEGE ADMISSION OFFICERS FOR SCHOOLS FOCUSED ON ROBOTICS KNOW THIS. THESE UNIVERSITIES HAVE A PREFERENCE FOR STUDENTS THAT HAVE COMPLETED A SEASON AS A MEMBER OF A ROBOTICS TEAM.

MANY COLLEGES ALSO BELIEVE THAT "A HEALTHY BODY EQUALS A HEALTHY MIND" AND LIKE TO SEE THEIR STUDENTS ACTIVE IN SOME WAY. EMPLOYERS OFTEN BELIEVE THE SAME. I ALSO ENCOURAGE MY STUDENTS TO MENTOR ELEMENTARY SCHOOL STUDENTS, AS I BELIEVE THAT IF A PERSON CAN TEACH A TOPIC, THEN THEY KNOW A TOPIC.

HOW CAN COLLEGE INTERNSHIPS PREPARE STUDENTS FOR FUTURE CAREERS IN ROBOTICS?

INTERNSHIPS CAN HELP FOCUS A STUDENT ON A DESIGNATED PATH IN ROBOTICS. ROBOTICS IS A WIDE FIELD, LIKE MEDICINE, AND INDIVIDUALS WILL NEED TO TAKE A MORE

FOCUSED PATH AS THEY PROGRESS IN THEIR STUDIES, AND ALSO WHEN IT'S TIME FOR THEM TO CHOOSE A JOB OR CAREER PATH. THEY CAN CHOOSE TO PURSUE ROBOTICS FROM A HARDWARE OR SOFTWARE ANGLE, OR FOCUS ON MILITARY DEVICES, EXPLORATORY DEVICES, MEDICAL DEVICES, ETC. INTERNSHIPS SHOULD BE USED AS A LEARNING EXPERIENCE AND AS A WAY TO NARROW ONE'S FOCUS.

WHAT IS THE OUTLOOK FOR ROBOTICS CAREERS?

I BELIEVE THE FUTURE OF ROBOTIC CAREERS WILL PARALLEL THE GROWTH OCCURRING IN THE ENGINEERING AND COMPUTER SCIENCE FIELDS. NEW CAREERS IN ROBOTICS WILL DEFINITELY BE CREATED. I GENERALLY TELL MY STUDENTS TO STUDY AT A UNIVERSITY THAT WILL ALLOW RESEARCH AND STUDIES IN THE NEWEST TECHNOLOGIES, AND TO LET THE WORK THEY DO DETERMINE THEIR CAREERS. THIS IS MORE EFFECTIVE THAN TRYING TO PURSUE A PREDETERMINED CAREER.

THE ADVICE THAT I GIVE TO MY STUDENTS IS, "TRY WHAT YOU ARE INTERESTED IN, BE SURE YOU ENJOY IT, AND HAVE A PASSION FOR SUCCESS."

GLOSSARY

dissertation A long paper, written in order to get a doctorate degree, that presents the results of a student's research.

doctorate An advanced degree that is earned after getting a master's degree.

drafting The act of creating formal plans or blueprints.

ethics The rules of moral conduct governing an individual or a group.

excavate To dig something up, or to dig a hole.

genetics The study of inherited traits.

humanoid Having human form or characteristics.

incision A cut made into the body during surgery.

internship A job performed, usually for little or no money, by students seeking to gain work experience.

locomotion The act of moving from place to place.

master's degree An advanced college degree, earned in one or more years of study after a bachelor's degree.

pharmacy A place where medicine is made or distributed.

postgraduate Relating to education undertaken after the completion of college.

prescription A written direction or order for the preparation and use of a medicine.

prostheses Artificial devices to replace a missing or nonworking part of a human body.

tournament A contest or series of contests played for a championship.

trait A quality or characteristic.

For More Information

FIRST
200 Bedford Street
Manchester, NH 03101
(800) 871-8326
Web site: http://www.usfirst.org
FIRST sponsors national robotics contests and is dedicated to getting young people interested in science and technology.

FIRST Robotics Canada
Richard Yasui
FIRST Robotics Administrator
Toronto District School Board
140 Borough Drive, Level 1
Toronto, ON M1P 4N6
Canada
(416) 396-5907
Web site: http://www.firstroboticscanada.org
FIRST Robotics Canada sponsors robotics contests in Canada.

Robotics Alliance Project
National Aeronautics and Space Administration (NASA)
NASA Headquarters
300 E Street SW
Washington, DC 20546-0001
(202) 358-0001
Web site: http://www.robotics.nasa.gov

The Robotics Alliance Project is a branch of NASA that focuses on getting students interested in and prepared for robotics careers in the space industry.

Skills/Compétences Canada (S/CC)
260 Saint Raymond Boulevard, Suite 205
Gatineau, QC J9A 3G7
Canada
(819) 771-7545
Web site: http://www.skillscanada.com
S/CC is a national, nonprofit organization that works with employers, educators, labor groups, and governments to promote skilled trades and technology careers for Canadian students.

WEB SITES

Due to the changing nature of Internet links, Rosen Publishing has developed an online list of Web sites related to the subject of this book. This site is updated regularly. Please use this link to access the list:

http://www.rosenlinks.com/robo/roca

FOR FURTHER READING

Cook, David. *Robot Building for Beginners.* 2nd ed. New York, NY: Apress, 2010.

Hyland, Tony. *High-Risk Robots.* North Mankato, MN: Smart Apple Media, 2008.

Hyland, Tony. *Robots at Work and Play.* North Mankato, MN: Smart Apple Media, 2008.

Hyland, Tony. *Scientific and Medical Robots.* North Mankato, MN: Smart Apple Media, 2008.

Hyland, Tony. *Space Robots.* North Mankato, MN: Smart Apple Media, 2008.

Jefferis, David. *Robot Brains.* New York, NY: Crabtree Publishing Company, 2007.

Jefferis, David. *Robot Warriors.* New York, NY: Crabtree Publishing Company, 2007.

Loy, Jessica. *When I Grow Up: A Young Person's Guide to Interesting and Unusual Occupations.* New York, NY: Henry Holt, 2008.

Manatt, Kathleen. *Cool Science Careers: Robot Scientist.* Ann Arbor, MI: Cherry Lake Publishing, 2008.

Piddock, Charles. *Future Tech: From Personal Robots to Motorized Monocycle.* Washington, DC: National Geographic Society, 2009.

Randolph, Ryan. *Robotics.* New York, NY: PowerKids Press, 2009.

Rhodes, Faye. *The LEGO Mindstorms NXT Zoo: A Kid-Friendly Guide to Building Animals with the NXT Robotics System.* San Francisco, CA: No Starch Press, 2008.

Steffoff, Rebecca. *Robots.* New York, NY: Marshall Cavendish, 2008.

BIBLIOGRAPHY

Boston Dynamics. "Big Dog: The Most Advanced Rough-Terrain Robot on Earth." 2009. Retrieved March 10, 2009 (http://www.bostondynamics.com/robot_big dog.html).

Cohn, Jessica. *Top Careers in Two Years: Manufacturing and Transportation.* New York, NY: Ferguson Publishing, 2008.

Ferguson Publishing. *Careers in Focus: Engineering.* New York, NY: Ferguson/Infobase Publishing, 2007.

Ferguson Publishing. *Discovering Careers for Your Future: Space Exploration.* New York, NY: Ferguson/Infobase Publishing, 2008.

Hyland, Tony. *Robots at Work and Play.* North Mankato, MN: Smart Apple Media, 2008.

Hyland, Tony. *Space Robots.* North Mankato, MN: Smart Apple Media, 2008.

Kupperberg, Paul. *Careers in Robotics.* New York, NY: Rosen Publishing Group, 2007.

Mataric, Maja M. *The Robotics Primer.* Cambridge, MA: MIT Press, 2007.

McGraw-Hill editors and the U.S. Department of Labor, Bureau of Labor Statistics. *The Big Book of Jobs.* 2009–2010 ed. New York, NY: McGraw-Hill, 2009.

NASA. "R2." January 11, 2010. Retrieved February 25, 2010 (http://robonaut.jsc.nasa.gov).

National Robotics Engineering Center. "Autonomous Loading System (ALS)." Retrieved March 10, 2010 (http://www.rec.ri.cmu.edu/projects/als).

O-NET OnLine. "Summary Report for: 17-3024.01—Robotics Technicians." 2009. Retrieved February 3,

2009 (http://online.onetcenter.org/link/summary/
17-3024.01).

Piddock, Charles. *Future Tech: From Personal Robots to
Motorized Monocycle.* Washington, DC: National
Geographic Society, 2009.

Rothman, Wilson. "Rescuer by Remote: Need Help?
Send in the Robot." June 8, 2004. Retrieved June 17,
2010 (http://www.time.com/time/2004/innovators/
200406/murphy.html).

Schneider, David. "Robin Murphy: Roboticist to the
Rescue." *IEEE Spectrum: Inside Technology*, February
2009. Retrieved February 2, 2009 (http://spectrum.
ieee.org/robotics/artificial-intelligence/robin-
murphy-roboticist-to-the-rescue).

Steffoff, Rebecca. *Robots.* New York, NY: Marshall
Cavendish, 2008.

Tech Museum of Innovation. "Machines and Man: Ethics
and Robotics in the 21st Century." 2005. Retrieved
February 3, 2009 (http://www.thetech.org/robotics/
ethics/index.html).

Trafton, Anne. "From Nature, Robots: Mechanical
Engineer Sangbae Kim Looks to Animals to Inspire
His Robot Designs." *MITnews*, September 25, 2009.
Retrieved March 10, 2009 (http://web.mit.edu/news
office/2009/stickybot-092509.html).

INDEX

ABOUT THE AUTHOR

Simone Payment has a degree in psychology from Cornell University and a master's degree in elementary education from Wheelock College. She is the author of twenty-four books for young adults. Her book *Inside Special Operations: Navy SEALs* (also from Rosen Publishing) won a 2004 Quick Picks for Reluctant Young Readers award from the American Library Association and is on the Nonfiction Honor List of Voice of Youth Advocates.

PHOTO CREDITS

Cover, p. 36 Yoshikazu Tsuno/AFP/Getty Images; back cover and interior © Axel Lauerer/Flickr/Getty Images; p. 8 Sebastian D'Souza/AFP/Getty Images; p. 13 U.S. Navy Photo by John F. Williams; p. 15 www.newtonium.com; p. 17 Courtesy of Donna Coveney, Massachusetts Institute of Technology; p. 24 John Nordell/Christian Science Monitor/Getty Images; p. 26 Carnegie Mellon University; p. 27 Shutterstock.com; p. 32 NASA Langley Research Center Collection/Sean Smith; p. 37 Gareth Evans, Aberystwyth University.

Designer: Matthew Cauli; Editor: Nicholas Croce; Photo Researcher: Peter Tomlinson